Edward Meryon

On the Functions of the sympathetic System of Nerves

As a physiological Basis for a rational System of Therapeutics

Edward Meryon

On the Functions of the sympathetic System of Nerves
As a physiological Basis for a rational System of Therapeutics

ISBN/EAN: 9783337139902

Printed in Europe, USA, Canada, Australia, Japan

Cover: Foto ©berggeist007 / pixelio.de

More available books at **www.hansebooks.com**

ON THE

FUNCTIONS OF THE

SYMPATHETIC SYSTEM OF NERVES,

AS A PHYSIOLOGICAL BASIS

FOR A

RATIONAL SYSTEM OF THERAPEUTICS.

By EDWARD MERYON, M.D., F.R.C.P.,

LATE LECTURER ON COMPARATIVE ANATOMY AT ST. THOMAS'S HOSPITAL.

LONDON:

J. & A. CHURCHILL, NEW BURLINGTON STREET.

1872.

PREFACE.

A GREAT portion of the following contribution towards a rational system of Therapeutics, was published in the *Lancet* during the months of October and November, 1871. Since that time I have received so many complimentary communications from my fellow-workmen, at home and abroad, that I venture to print it in a separate form, with the hope that it may assist, be it in never so small a degree, to establish therapeutics on a scientific and rational basis. To render it perfectly intelligible, I have been obliged to supply some prefatory observations on the structure and functions of the sympathetic system of nerves, for on such anatomical and physiological facts my theory is founded. The anatomical part of the subject I have carefully pursued myself, and have seldom failed to trace three kinds of nerve-fibres to every sympathetic ganglion that I have examined. The physiological experiments I have culled from sources which, I trust, will be accepted as perfectly trustworthy.

FIRST PART.

ON THE STRUCTURE OF THE SYMPATHETIC SYSTEM OF NERVES.

THE studious research which has occupied the minds of many physiologists of late years, relative to the special functions of the sympathetic system of nerves, has induced me to enter the list of inquirers; and although I can only claim, as a qualification for the task I have undertaken, a somewhat extensive course of dissection, not only of the human subject but of the lower animals, together with the possession of numerous facts accumulated by the observations of others, and which I have in common with every physiologist, I venture to offer such generalisations as will, I trust, lead to more positive knowledge than we now possess.

In the first place, I have long thought that there is great significance in the fact that every sympathetic ganglion is connected with both motor and sensory nerves, as well as with its own special nerve-fibres (the so-called nerves of Remak).

Thus the superior cervical ganglion, in addition to the fibres of the sympathetic proper, receives branches from the three or four upper cervical nerves; one branch from the hypoglossal, one from the pneumogastric, and one from the glosso-pharyngeal.

Its affiliated ganglion, the petrosal, has a branch given off to it from the facial nerve, and one from the glosso-pharyngeal.

When the carotid ganglion exists in the cavernous sinus, it invariably receives a branch from the sixth nerve, and one from the vidian nerve, proceeding from Meckel's ganglion.

The lenticular ganglion has a communicating branch from the motor oculi (the third), and one from the nasal branch of the ophthalmic nerve.

The geniculate ganglion (an expansion of the portio intermedia of the facial nerve) is connected with the vidian and facial nerves, with the spheno palatine branch of the fifth, and with the lenticular and otic ganglia.

The otic ganglion receives a branch from the motor branch of the inferior maxillary nerve, one from the facial, and one from the glosso-pharyngeal.

The spheno-palatine ganglion, or the ganglion of Meckel, communicates with the facial nerve, through the intervention of the vidian nerve, and with the superior maxillary nerve.

The submaxillary ganglion has a communicating branch from the hypoglossal and chorda-tympani combined, and one from the gustatory nerve.

The middle cervical ganglion receives combined

motor and sensory nerves from the second, third, fourth, fifth, and sixth cervical nerves.

The inferior cervical ganglion has also communicating with it branches from the seventh and eighth cervical and first dorsal nerves; it acquires branches likewise from the phrenic and lower laryngeal or recurrent nerves, the latter consisting of sensory fibres from the vagus.

The cardiac ganglion is formed by the cardiac nerves from opposite sides, which contain fibres from the cerebro-spinal system; also by branches from the main trunk, and from the recurrent division of the pneumogastric.

Each of the thoracic ganglia receives two communicating branches from the corresponding spinal nerves.

The semilunar ganglia are formed chiefly by the great splanchnic nerves, the phrenic, and the pneumogastric nerves.

The renal ganglia receive compound branches from the lesser splanchnic and the lumbar nerves.

The lumbar ganglia have each two communicating branches from the spinal nerves.

Finally, each of the four pairs of sacral ganglia, and the terminal ganglion on the coccyx, receives two communicating branches from the spinal nerves.

The same combination of motor and sensory nerve-fibres obtains in every plexus formed by the sympathetic system. They are not very abundant, however, in the mesenteric plexuses, but they exist nevertheless. In the hypogastric plexuses they abound, as they do in the parotidean plexus, likewise in the inferior mesenteric and uterine, which receive two or three branches each from the sacral nerves.

It will be observed that in many instances more than one nerve of the same kind—either motor or sensory—go to a single ganglion, and it is doubtless for the purpose of conveying as many influences, and not for the purpose of accumulating the same kind of nervous activity. This proposition has been strongly insisted on by M. Claude Bernard.

An interchange of fibres takes place between each pair of ganglia in the main cords of the sympathetic system and the corresponding spinal nerves. White nerve-fibres proceed from the spinal cord to the ganglia, and grey nucleated fibres proceed from the ganglia to the spinal cord; and it has been shown by Kölliker that in man and the higher animals, more grey fibres pass from the ganglia to the cord than white fibres from the cord to the ganglia. He also confirmed an observation made by Dr. Beck, that the grey fibres give off minute branches to the bloodvessels in their passage to the cord: The remaining branches have been traced by Müller both to the anterior and posterior roots of the spinal nerves.

It is this interchange of fibres, probably, which determines the position of the principal sympathetic ganglia in the immediate vicinity of the spinal cord.

Each ganglion consists of cellular and fibriform substances, surrounded by a comparatively dense tissue, which is a continuation of the so-called sheath of Schwann.

The structure of the ganglionic cells has been

carefully studied and described by Dr. Lionel Beale as possessing the same general character in all animals.*

The substance of the cell consists of more or less granular material, and near the fundus there is a large circular nucleus, with its nucleolus. The size of the cell, with its investing membrane, which is also a continuation of the sheath of Schwann, is about the 0·060 of a millimeter in diameter; the granular matter may have a diameter of about 0·030 of a millimeter; the nucleus of about 0·012; and the nucleolus of 0·002.

In the centre of each cell the granular matter gradually assumes the form of a nerve-fibre, which projects to constitute a process or pole of the cell, and is continued outwards in the form of what Dr. Lionel Beale has called the straight fibre. At the circumference of the cell the granular matter also assumes the form of a nerve-fibre, and projects in close proximity with the pole of the straight fibre, around which it winds in a spiral manner. These latter, or spiral fibres, according to Beale, after having surrounded the straight fibres, are continued in a direction parallel with them for a short distance, but eventually turn and take a course diametrically contrary to that taken by the straight fibres. Both straight and spiral fibrils can be shown to be continuous with the granular matter of which the body of the cell is composed.

In the sympathetic ganglia, both multipolar and unipolar cells exist, each cell appearing to be a mere enlargement of the axis-cylinder of the nerve. In

* Philosophical Transactions, 1863, vol. cliii. p. 539.

other words, each single cell is the origin or starting point of one or more nerve-fibres; for as a cell may contain many nuclei and nucleoli,* each nucleus originating a primitive fibre, as many fibres may start from that cell, be it unipolar or multipolar, and run into a bundle of other fibres proceeding from other cells.

On emerging from the ganglia the nerve-fibres have a very complicated and intricate arrangement; but it is enough for our purpose to be certain of the continuity of the several nerve-fibres with the unipolar or multipolar cells. Dr. Beale, in the paper above referred to, asserts that nerve-cells are always connected with nerve-fibres; and Max Schultz endorses the fact by the statement that the processes of cells are nerve-fibres, as was first observed by Remak in the vertebrata, and by Helmholtz among the invertebrata.†

Valentin and Bidder originated a doubt relative to the nature of the so-called grey fibres; Bidder supposed them to be a variety of areolar tissue, but they are now generally accepted as nerves, and are recognised as such under the several names of nuclear fibres, fibres of Remak, non-medullated sympathetic fibres of Max Schultz, and gelatinous fibres of Henle.

Remak, in his essay on multipolar cells, maintains that these bodies in the main cords of the sympathetic become continuous, by means of their caudate processes, with the axis-cylinders of both the broad

---

* Human and Comparative Histology, by Stricker, translated by Mr. H. Power for the New Sydenham Society.—General Characters of Nervous Tissue, by Max Schultz, p. 175.

† Ibid. p. 174.

and fine cylindrical white nerve-fibres, and with the gelatinous grey fibres. It may, therefore, be affirmed that every form of nerve-fibre in the ganglia is connected and continuous with the ganglionic cells.

Every ganglion thus possesses all the elements of a nervous centre, and the researches of physiologists tend to confirm the inference that such, in effect, is the case; and that each, in its own sphere, is capable of receiving, transmitting, originating, and reflecting impressions on which the healthy functions of the organs to which its nerves are sent depend.

The question relative to the manner in which nerves terminate is an exceedingly interesting one; and if the sympathetic nerve-fibres could be unequivocally traced to the special tissues wherein they end, a fair inference might be drawn as to their function. Some have been followed to their peripheral extremities by Dr. L. Beale, and he has described them as distributed over the walls of vessels. Dr. Tyson, of Pennsylvania, has endorsed his views; and Professor Eberth, of Zurich, has demonstrated the presence of nerves in the coats of all vessels, the capillaries excepted, even in the tunica adventitia of the non-muscular veins of the pia mater. And these, be it observed, *partly consisting of dark-edged, and partly of pale fibres*, which break up after they have penetrated the tunica adventitia into a fine net-work.*

There can be little doubt, however, that the sympathetic nerves also stand in intimate relation to the

* Stricker, op. cit., vol. i. p. 266.

secreting cells of glands. Their distribution in the salivary gland has been described by Pflüger.* In this organ the dark-edged or medullated nerves constitute the greater number and accompany the salivary tubes, perforate the membrana propria, then divide into innumerable fibrils, each of which becomes continuous with a salivary cell. The pale or non-medullated nerves are composed of extraordinarily fine fibrils, each of which is continuous with the fibrillated substance of the epithelial cells.

The axis-cylinder which invests these pale fibres, is found to be continuous with the membrana propria.

Such being the anatomical relationship subsisting in the so-called sympathetic system of nerves, much circumspection is necessary in drawing conclusions from experiments, seeing that the cerebro-spinal nerves are intimately commingled with the nerves of Remak, all running parallel to each other.

I may observe, incidentally, that such compound nerves appear to fulfil all the conditions which are said to be necessary for what is called electro-tonus (a state whereby one nerve is rendered active by the activity of another in close proximity with it), and many phenomena which I may have to adduce seem to point to such influence; but I shall endeavour to confine my remarks to such experiments and observations as have reference to single nerves only.

This, however, can be done in no other way than by comparing the phenomena induced and presented by the cerebro-spinal nerves alone, with such other phenomena as are produced by the special agency of

* Stricker, vol. i. pp. 433—448.

the fibres of Remak, and by investigating in what respect the mode of action of the one set differs from that of the other; for we may assume that Nature is too good an economist to endow the fibres of one with attributes possessed by the other, when included in the same nerve.

It is very probable that all may be in some degree modified, and it is difficult to conceive why the ganglionic cells are the connecting media between the three sets of nerves, if some change of function be not the result. In illustration of this, the sensory nerves appear to lose much of their peculiar sensibility, as we recognise it in their ordinary impressions on the brain, when they proceed from an organ to a ganglionic centre of the sympathetic system; and if so, it follows that, by virtue of the reciprocal action which sensitive and motor nerves exercise on each other, and on the organs to which they are distributed, the motor fibres must be invested with properties derived from the modified sensitive nerves.

What, then, is the special function of each different form of nerve-fibre respectively which goes to or proceeds from every ganglionic centre?

With the view to an explicit answer to this question, the most obvious method is to select from the three forms of nerve-fibre the single one which is invariably associated with a peculiar phenomenon, and without which that phenomenon does not occur. The logical "method of difference," in effect, is that which I shall endeavour to pursue.

ON THE FUNCTIONS OF THE SYMPATHETIC SYSTEM OF
NERVES.

In M. Claude Bernard's experiments on the sympathetic nerves, the results of which he communicated to the French Academy,* the interesting fact
was educed, that whenever the sympathetic nerve
in the neck of a rabbit is divided, an elevation of
temperature occurs in the tissues on the corresponding side of the head, amounting to 7° Fahr. when
contrasted with the uninjured side. This increase
of heat was plainly perceptible by the hand, and
admitted of accurate measurement by the introduc-
tion of the bulb of a thermometer within the nostril,
or into the external auditory meatus. The whole
body shared, to a certain extent, in this develop-
ment of heat, and exhibited evidence of a temperature exceeding the natural standard; but it was
most evident on the side of the neck where the
sympathetic had been divided, and least so on the
opposite side where it had been left uninjured. The
mercury rose to 72° Fahr. on the affected side, but
only to 68° on the uninjured side. Nor was this
elevation a transitory phenomenon, for it continued
with remarkable steadiness until the animal was
killed, and even after death the side of the neck on
which the experiment was practised was the last
part of the body to lose its vital heat. In some
other cases the increased heat disappeared, but in

* Comptes Rendus, vol. xxxiv. p. 472. Fevrier, 1852.

no instance was there œdema, nor any morbid phenomenon resembling inflammation.

In addition to the above phenomena, M. Bernard
subsequently noticed an increased temperature of
the cerebral hemispheres, as well as of the blood
itself in the internal jugular vein on the side in
which the sympathetic nerve had been divided.[*]

Schiff has repeated these experiments, and infers,
from precisely the same results, that *active dilatation*
is a function possessed by blood vessels.

Dr. W. Ogle communicated to the Royal Medical
and Chirurgical Society the history of a suppurating
tumour in the neck of a man, which produced a
lesion equivalent to a division of the cervical sympathetic.[†] In this case the ear on the affected side
was redder and warmer by two degrees Fahrenheit,
than that on the opposite side, and there was a
total cessation of cutaneous secretion on the right
(*affected*) side of the face, head, and neck, although
the skin of the right cheek was pinker than that of
the left.

We may infer, then, that the disseverance of the
minute arteries from the influence of the nerves of
Remak is productive of increased vascularity, and
an elevation of the temperature of the parts so dis-
connected.

M. Claude Bernard exposed in a dog the gustatory nerve, the chorda tympani (before they receive
communicating branches from the lingual nerve),
and the submaxillary ganglion. Having thus before
him a sensory nerve, a motor nerve, and a ganglio-

---

* Mémoires de la Société de Biologie. 1853.
† Medico-Chirurgical Transactions, vol. lii. p. 151. 1869.

nic centre, he divided the gustatory nerve, and the secretion from the gland was immediately stopped. He then pinched the centripetal end of the cut nerve which communicates with the brain, and a large quantity of saliva was secreted, whilst the ducts of the parotid and sublingual glands remained dry.*. He subsequently varied the experiment by cutting the chorda tympani, leaving the gustatory nerve intact, and secretion immediately ceased, as in the first experiment. He then inserted a tube into the Whartonian duct, and communicated a weak current to the peripheric end of the divided nerve. Every time that this was done, a drop of saliva was seen to fall from the tube. Thus secretion was arrested by section of the tympano-lingual nerve, and reproduced by the stimulus of electricity communicated to its distal extremity. When applied to the centripetal end of the divided nerve, the electric current had no effect.

These experiments supply us with something like evidence, from which to infer that secretion may be, in some way or other, dependent on cerebro-spinal influence, modified, as I have before suggested ; and as the nerve-current of a sensitive nerve, on which the first experiment was performed, is centripetal, the irritation of pinching the central portion of the cut nerve could only be conveyed to the gland by the reflex action of the returning motor nerve. Whereas when the chorda tympani was divided, secretion was re-established by communicating an electric current through the peripheric end of the nerve. But the electric current has no effect when

* Comptes Rendus, vol. xxxiv. p. 474.

applied to the peripheric portion of a divided sensory nerve; and we are therefore warranted in the assumption that it is through the influence of the motor nerve that secretion is re-established.

Something more, however, is suggested to the mind by the fact of the ducts of the parotid and sublingual glands remaining dry, whilst the Whartonian duct poured out saliva; and that is, that every gland stands in relation to a special act, and that its function is determined by a special and independent influence.

If the facial nerve be divided as it passes out of the stylo-mastoid foramen, the secretion of the parotid gland is but little affected. If, however, the facial be divided at its origin, inside the skull, secretion, both from the parotid and submaxillary glands, is abolished.* There is something, therefore, between the origin of the facial nerve and the stylo-mastoid foramen, to which such arrest of secretion is due. It cannot be the facial alone, for its ablation has but little effect, either on the parotid or submaxillary gland. Neither can it be the chorda tympani alone, for section of that nerve does not influence the parotid secretion, but stops that of the submaxillary. But the division of the facial nerve in the hiatus Fallopii involves the resection of the portio intermedia of Wrisberg, which expands, in the Fallopian canal, into the geniculate ganglion. Thus we have an indication that the secretion of the parotid is dependent on a ganglionic centre.

If the facial nerve be left uninjured, and the nerves

* Leçons sur la Physiologie et la Pathologie du Système Nervaux. M. C. Bernard. Tome ii. p. 154, 155.

of Remak (which proceed from the superior cervical ganglion to be distributed to the ramifications of the internal maxillary artery) are cut, the secretion of saliva goes on more abundantly and continuously. It has also been observed that, when the floor of the fourth ventricle is slightly wounded in the immediate vicinity of the nuclei of the fifth pair of nerves, well-marked salivation from the parotid glands is set up. If the wound be in the mesial line, ptyalism is induced on both sides; but if the wound be on one side, then the increased secretion is occasioned in the parotid of the opposite side.*

The results of these experiments are strictly in accord with those on which I have already commented; arrested secretion by section of the motor fibres of the vaso-motor nerves, increased secretion by section of the fibres of Remak, and increased secretion by excitation of the sensory nerves.

M. Claude Bernard divided the sympathetic in the upper part of the dorsal region of a horse; a greatly increased vascularity was the immediate result, and the corresponding parts of the surface were bathed in sweat.†

He also found considerable distention of the pericardial vessels, and serous exudation from them, after injuring the cardiac ganglia of the sympathetic. This experiment was repeated and verified by Schiff, and Remak explained the phenomena by the assumption that when the bloodvessels are deprived of the inhibitory influence of the sympathetic nerves proper, they dilate and allow blood-corpuscles to

* *Medical Times and Gazette,* 1860, p. 362.
† Ibid. 1861, p. 544.

penetrate into those minute arterioles, through which blood-plasma only should be propelled.*

On injuring the solar plexus, or on dividing the main trunks of the sympathetic, Budge ascertained that the circulation of the blood in the liver is increased, and the secretion of bile augmented, and in two cases he found the liver itself enlarged. On extirpating the mesenteric plexuses in rabbits, the fæcal pellets became so soft that none of the ordinary rounded masses were found in the rectum. The fæces were pulpy, and covered over with a slimy mucus.† M. Claude Bernard extirpated the semilunar ganglia in a large shepherd's dog, and observed the same results.‡

Jaschkowitz, by dividing the sympathetic nerves of the spleen in cats and dogs, caused an increased flow of blood to that organ, and a copious deposit of hæmatin pigment in its cells.§

In the experiments instituted by Bernstein on the pancreas, he discovered that, notwithstanding the absence of secretion in that organ during fasting, on dividing the sympathetic nerves going to it, a continuous flow of pancreatic fluid is produced.‖ The originating power of the ganglionic centres is

* The ulterior stages of inflammation have been still further explained by Waller, Reicher, Cohnheim, and Stricker, by the proposition that pus-corpuscles are nothing but the colourless blood-corpuscles filtered from the blood through the walls of capillary vessels. In a very elaborate paper contained in Virchow's *Archiv,* for September, 1867, Cohnheim demonstrates the truth of this proposition by exciting inflammation in the peritoneum of frogs.

† Nova Acta, Acad. Cæs. Leop. Car. Nat. Cur. xix. p. 257, 1860.

‡ Leçons sur la Physiologie et la Pathologie du Système Nervaux. Tome ii. p. 522.

§ Jaschkowitz de Discisionis Plexus Lienalis efficitate in Lienem.

‖ Sächs Akad. Sitzungberichte Math. Phys. Class, 1869.

indicated by the fact, that when secretion is thus artificially produced, it is not arrested by placing the animal under the influence of woorara. The special relation, moreover, in which the action of the pancreas stands to the digestion of food, increasing immediately after a meal, attaining its maximum about three hours after, and then gradually diminishing,* shows the local centralisation of nervous influence.

The structure of the kidneys is almost suggestive of the function of each component part.

The Malpighian tufts, and the special capillaries surrounding them, shut up within the Malpighian capsules, in which there is but little of epithelial cell structure, exhibit the most perfect contrivance for that simple and independent filtration of fluid which is known as exosmose—a simple outpouring of the watery part of the blood into the capsules, to flush, as it were, the tortuous tubuli uriniferi in front.

The cortorted tubuli, surrounded by *their* special capillaries, to which the ultimate fibres of the renal plexus have been traced, and constructed of basement membrane, lined with the rounded, glandular variety of epithelial cells, indicate a secretory structure; and here, undoubtedly, it is that the salts which characterise the urine are eliminated from the blood.

Section of the splanchnic nerves, as performed by Eckhardt, induces hyperæmia of these latter capillaries, albuminuria, and increased renal secretion;

* Arbeiten aus der Physiologischen Anstaltzen. Leipzig, Vierter Jahrgang, 1869.

and Dr. James Tyson, of Pennsylvania, has shown that by section or stimulation of the gangliated nerve-fibres, the terminations of which he has traced to the minute vessels surrounding the contorted tubuli uriniferi, he produced either distended vessels and an abundant secretion, or a diminished and slower blood-current, and a corresponding secretion.

The question as to which of the three forms of nerve-fibres it is that is subservient to the special function of the secretory tissues, admits of a positive answer. It cannot be the fibres of Remak, because when they are cut, secretion is increased; neither can it be the sensory fibres, for they are centripetal nerves, and their influence can only be of a reflex character. The motor fibres are those only which, going to a gland, cannot be cut without destroying secretion; and those fibres, therefore, must be the cause or condition of the phenomenon. The sensory nerve-fibres may endow the secretory cells with their vital sense, as exemplified in Bernstein's experiments on the pancreas; but the motor fibres must be immediately concerned in supplying blood for increased secretion.

The following case is illustrative of my inference: R. de L——, aged forty-three, presented herself at the Infirmary for Epilepsy and Paralysis, in July, 1869, complaining of spasm of the muscles supplied by the facial nerve on the left side. Some two years before she had a severe headache, after which she awoke one morning with her mouth drawn to the left side, with a feeling of stiffness, but no anæsthesia. From that time any extraordinary action of the muscles of the face induces a contraction of the buccinator, masseter, digastric, stylohyoid, and orbicularis palpebrarum muscles. The act of sneezing, gaping, or blowing her nose, will produce this spasm, which continues for five or six minutes, when the face regains its tranquil appearance. She has occasional dyspnœa, *a copious flow of saliva from the left parotid gland,* and the left side of the face and head often sweats when the right does not.

The result of all similar experiments upon every secreting organ in the animal body, is such as I have described, and the inevitable inference is, that without the motor nerve-fibre of the vaso-motor nerves, secretion does not take place.

Let us now inquire—What is the agency or influence of the sensory nerve-fibres in the sympathetic system?

When any pungent substance, such as mustard, is placed in the mouth, it produces lacrymation, and an increased flow of saliva. Such increment of secretion results from the application of any excitant to the extremities of the sensitive nerves, which emanate from glandular organs. Professor Ludwig has affirmed that the augmented exudation is owing neither to contraction of the muscular elements of the glands, nor to a change in the pressure of blood in the organs, but apparently to the simple influence on the secreting tissues; that a stimulus to secretion is set up, not by any direct influence on the blood-vessels, but by an incitement set up in the secreting glandular tissue. This may be a refinement on a physiological definition; for it is difficult to conceive how secretion can be effected without a corresponding increase of vascular action, but it is the expression of an ideal reciprocation of tissue with tissue, which is suggestive of spontaneity. A reciprocity, according to Stilling, between sensory and vaso-motor nerves.

The act of blushing is an apposite instance of the correlation of sensitive and vaso-motor influences. But paralysis of the sensory nerves, although it may diminish, does not stop vascular action; or, on the occurrence of anæsthesia, gangrene should supervene in the parts deprived of sensation. The tendency to such destruction may exist by virtue of the response given by the motor nerves to those which are rendered insensible, and, therefore, of the loss of balance between the incentive and restraining forces which act upon the bloodvessels. More than forty years ago M. Majendie observed that after division of the trigeminus in reptiles, which resist the fatal influence longer than do the mammalia, the side of the face on which the section is made, *did* actually become gangrenous.

Paralysis of the trigeminus, however induced, causes anæsthesia of the face and scalp, injection of the conjunctivæ, loss of secretion of the lacrymal glands, leucoma, and ulceration of the corneæ. A case of this kind is reported by Dr. Althaus, in the fifty-second volume of the "Medico-Chirurgical Transactions." In that instance there was also a continuous rushing noise in the head, an abundant discharge of acrid mucus from the nose, and a copious buccal exudation, although the secretion of healthy saliva was arrested.

It is a curious fact, and a very interesting one

that the section of a sensitive nerve is very apt to be followed by a pseudo inflammation in the parts to which the peripheral extremities of that nerve are distributed. Professor Graefe has abundantly proved that such apparent inflammation in distant parts is very apt to follow the section of a sensory nerve, by which those parts are supplied. He asserts, that in operations on the eye, inflammation is not so apt to occur when the ophthalmic nerve is left uninjured, as when it is cut. According to the view here taken of the special function of the sensory sympathetic fibres, it is reasonable to suppose that the phenomenon is due to a loss of tonicity of the walls of the capillary vessels, owing to a suspension of the inhibitory influence of the vaso-motor nerves, and that blood-corpuscles may thus penetrate into minute vessels, where only blood plasma, nutritive fluid, or effete matter should be carried to the surface.

It is this peculiar nerve influence, modifying indirectly the nutrition of organs, which has given rise to the idea of trophic nerves; and in all cases the nerves so implicated are sensory nerves. Their action is exemplified in gastrodynia and dysmenorrhœa, in which they induce an increased flow of mucus, but in both diseases the pain is referable to nervous irritation, and independent, as Ludwig would have it, of vascularity of the secreting membranes.

Many instances are on record in which the so-called trophic nerves have manifested their influence on the nutrition of parts after accidents, by which the sensory nerves of those parts have been divided

or destroyed. Three cases of the kind are related
by Mr. Simon in Holmes's *System of Surgery*,* and
one has been described by Mr. Nunn,† in which the
ulnar nerve was divided in a boy. The temperature of the hand fell to 10° Fahrenheit below that
of the sound side, and the muscles supplied by the
divided nerve were wasted in the course of two
months to one-third of their former bulk. The
London Hospital Reports for 1866 contain the
details of seven cases of a like nature, and Dr. John
Ogle describes a case of disease in which the sympathetic must have been implicated, and a changed
condition of the vital attributes in the parts ensued,‡
which, as in all the other instances, manifested itself
a long time after by a lowered temperature, a
defective nutrition, and a disposition to textural
degeneration.

I adduce these cases to illustrate the apparent
paradox of correlations, yet dissociation of cerebro-
spinal and vaso-motor nerves.

All these phenomena are necessarily induced by
reflex action, either through the cerebro-spinal axis
or the sympathetic ganglia; and according as the
afferent nerve is excited or rendered apathetic, may
it either exalt or depress the functions of the nervous centres upon which it acts. On this theory
Dr. Handfield Jones and Mr. Lister§ explain the
distant effects of morbid changes.

In the foregoing experiments, the effect of section

---

* Vol. i. p. 36. Second Edition.
† Pathological Transactions, 1866.
‡ Medico-Chirurgical Transactions, vol. xli. 21st case.
§ *British Medical Journal*, February 5th, 1859.

of the fibres of Remak has been shown, in order to judge of the resulting action of the motor and sensory elements of the sympathetic system; and it is obvious that in all cases the attributes of the sympathetic proper are the very antitheses to those of the cerebro-spinal nerves. But as the sensitive element exerts its immediate influence on the nervous centres only, and the motor portion of the ganglionic centres on the peripheric tissues, it may be asserted that the latter and the fibres of Remak are antagonistic in their action.

Accordingly it is found that the section of the motor nerve of a gland produces precisely the same effects as stimulation of the corresponding sympathetic branch; and, *vice versâ*, that section of the sympathetic has the same effect as stimulation of the motor branch, both pointing to the fact which M. Claude Bernard has enunciated—that the nervous filament which presides over the functions of a gland is always from a motor nerve.*

I infer, therefore, that the special function of the sensory fibres of the sympathetic is to communicate a stimulus to the secretory glandular tissue, to give an organic or vital sense to those tissues, just as the muscular sense is conveyed from the muscles to the nervous centres to communicate a stimulus to muscular action. In other words, it is through the indirect influence of the sensory nerves that the vital phenomenon of secretion is induced, but by the direct operation of the motor-nerve fibres.

To what particular tissues, then, and to what

* *Medical Times and Gazette*, 1861, p. 645. Lectures on the Spinal Cord.

special function, are the sympathetic nerves *proper*, or the fibres of Remak, subservient ?

Their actions are undoubtedly complex, yet I hope to show (and I cannot discover from physiologists that I am mistaken) that those actions are confined to the muscular coats of vessels only, and that their function is chiefly to control the act of nutrition.

I have had occasion to refer to the researches of Dr. J. Tyson on the modes of distribution of nerves to minute arteries, and he has shown how the primitive fibrillæ of nucleated nerves are disposed upon the arterioles, which are distinguished by transverse nuclear markings—probably the contractile elements—and how other strands of naked axial fibres, on the uriniferous tubes of the kidneys, gradually become surrounded with a medullary tunic, and then with the tunic of Schwann, to constitute nerve-fibres which he believes to be sensory or afferent nerves, conveying to the centres an impression, the response to which is transmitted to the efferent nerves, or those which terminate on the transverse marks of the arterioles, whereby they contract or dilate their calibre. Such is the probable disposition of the grey fibres on the coats of vessels.

As before observed, I have anticipated experiments on the sympathetic fibres to show the behaviour of the motor fibres without them ; and in every instance in which the former have been divided, an increased vascularity, an elevated temperature, and an augmented secretion have resulted.

Another curious and instructive phenomenon occurs when the blood is thus transmitted in a preternatural quantity through the capillaries—the venous blood immediately becomes brighter in colour. M. C. Bernard observed this fact in the coronary veins on the left side of the lip of a horse after he had divided the left cervical sympathetic.[*]

Now, the application of a weak electric current to the peripheric end of the divided sympathetic reverses all this. The calibre of the distended capillaries is quickly reduced; the temperature is lowered, and may be depressed below the existing degree in other parts, and secretion is diminished. If the power of the current be increased, the circulation may be entirely arrested; so that, if examined under a microscope, the capillary vessels will seem to be completely empty.[†] Such is the invariable result of stimulation of the nerves of Remak on the capillaries; and MM. Valentin, Henle, and Budge have observed that the large blood-vessels contract, when acted on by galvanism, through the medium of the grey nerve-fibres which are supplied to them.

It appears, therefore, that all the conditions of healthy circulation and secretion are fulfilled in the reciprocal action of the three forms of nerve-fibres.

Make a section of them all, or cut away—say the renal of plexus—and all secretion of urine is arrested.

Increase the relative power of the motor nerve-fibres, by section of the nerves of Remak, and you establish a hyperæmia round about the Malpighian tufts, and diuresis.

Increase the relative power of the nerves of

* La Clinique Europ., 1859; No. xxix. p. 282.
† *Medical Times and Gazette*, 1861, p. 543: M. Claude Bernard on the Spinal Cord.

Remak, by section of the motor fibres, and you diminish circulation and secretion.

Increase the relative balance of power of either motor or inhibitory nerve-fibres, either by gently exciting the sensory nerves, or by painfully irritating them, and you have, in the first case, increased circulation and secretion, or, in the second case, the very reverse.

The results of these experiments have their analogues in the phenomena which we observe in a case of cholera.

During the stage of collapse there is a paralysed condition of the whole nervous system and suppression of urine, just as when section is made of the renal plexus.

As the motor fibres of the ganglionic centres recover power, lacking, however, the influence of the sensory elements, they do not induce a healthy secretion of the salts of the urine, but allow the escape of albumen from the blood—the first phenomenon, according to Schönbein, observed in diabetes—as when the nuclei of the pneumogastric nerves in the medulla oblongata are irritated.

By degrees the sensory nerves recover power, and their influence is transmitted to the epithelial cells; imperfectly at first—for as yet they are unchecked by the nerves of Remak, and sugar is poured out with, perhaps, a small portion of urea, just as was observed by Eckhardt, when he divided the splanchnic nerves, from which the abdominal ganglionic centres chiefly derive their fibres. As sugar appears albumen disappears. The slight impulse, also, given by the sensory nerves increases the secretion of the aqueous component of the urine, just as it is increased in diabetes.

At last the restraining nerve power of the sympathetic proper is regained; when, as above stated, all the conditions of healthy circulation and secretion are re-established. The sugar then gives place to urea, and the healthy secretion of urine.

If all this be so, we have at once a solution of very many physiological and pathological phenomena, and, what is more, a clue to the rational treatment or many diseases.

Mr. Moore, in his interesting little treatise " On Going to Sleep," shows the one way only in which sleep can occur; and *that*, the action of the sympathetic fibres in diminishing the arterial current through the brain. One step more and we have the key to the convulsions of teething; for MM. Schiff and Loven have shown that by irritating the peripheric extremities of sensitive nerves a reflex influence is produced on the vaso-motor nerves, inducing contraction of the capillaries;* and, as MM. Kussmaul and Tenner have explained, convulsions.† They have demonstrated, moreover, that a suddenly induced anæmic state of the brain produces epilepsy; and how frequently do both convulsions and epilepsy occur during sleep! Yet another step in the arrest of circulation, and we have the pathological state which M. Eckhardt has produced experimentally—namely, an over-irritated condition of the sensitive nerves reflected on the

---

* Ludwig's Arbeit, 1867.

† Nature and Origin of Epileptiform Convulsions. By Drs. Kussmaul and Tenner (New Sydenham Society, 1859).

vaso-motor nerves of the spine, and reflex paralysis as a consequence. It is M. Claude Bernard's experiment on the sympathetic over again; and I believe this to be the most frequent, if not the only, immediate cause of infantile paralysis, which, as far as I have been able to judge, is more relieved by the application of warmth, *as a local remedy*, to the spine, than by the stimulus of electricity.

In the above forms of convulsions, Dr. Marshall Hall's idea, that when the influence of the cerebrum is suddenly removed the augmented irritability is reflected back on the vaso-motor nerves and muscles, would appear to receive confirmation.

Hemicrania is said by M. Du-Bois Reymond, who has studied it on his own person, to arise from spasm of the muscular coats of the arteries of the brain; and M. Mölendorff, although not agreeing altogether with Du-Bois Reymond, considers that it is dependent on an affection of the cranial vaso-motor nerves.

It is a well-known fact that hemicrania is often attended with disturbance of vision, and it is also known that disturbed vision is sometimes the result of a diminished supply of blood to the brain; and that nervous headache very often follows hæmorrhage or other causes which induce a defective supply of blood to the entire brain, or to one side of it, and what more likely than spasm of the muscular coats of the arteries?

Now the active state of organs is unquestionably the wasting state, and the quiescent condition that of restoration. During mental activity the brain is charged with arterial blood, and in the glands the active state is a state of congestion. During this period the venous blood flowing from those organs is of a bright red colour, just as was observed by M. Bernard in the venous blood of the left coronary vein in the lip of the horse, after section of the left cervical sympathetic.

The period of sleep is regarded by Mr. Durham* and Mr. Moore as the period of replenishing the brain, and during the state of quiescence in glands those organs recuperate themselves. Then it is that the cerebro-spinal system is in abeyance, the fibres of Remak dominate, and the venous blood flowing from the organs is very dark in colour.

Thus the brain is subject to the common law which regulates the circulation of blood in other organs. When their function is excited the circulation of blood is increased, when they are quiescent the circulation is less active ; and the independence of the local circulation of blood in the cerebro-spinal centres has been accounted for by the presumed ganglionic character of the arachnoid membrane of the brain and spinal marrow.†

In the face of all these phenomena it cannot be doubted that nutrition chiefly is governed by the sympathetic nerves, whilst local independence is indicated by the local distribution of ganglionic centres, and proved by local vascularity during functional activity.

In illustration of these facts, I would observe that reflex actions of vaso-motor nerves are limited to the

* Guy's Hospital Reports, 1860 : The Physiology of Sleep.

† Medico-Chirurgical Transactions, vol. xxix., p. 581: paper by G. Rainey, Esq.

particular organs or parts supplied by those nerves; whilst the reflex actions of ordinary sensitive nerves are manifested on both sides of the body; thus showing that the first set of phenomena have their centres in the sympathetic ganglia, and the latter in the spinal cord.

When the salivary glands are stimulated by the act of masticating and by the savour of food, the blood-vessels of the glands become turgid, and the glands themselves injected. The contact of food quickens the circulation of the mucous membrane of the stomach, and causes it to become red and tumefied. Again,—during menstruation, a period of nervous sensibility, the whole uterus is congested; the mucous lining throughout its whole extent is hypertrophied and of a pink colour, increasing in depth towards its fundus, where it attains a dark livid colour, and is covered over with a quantity of blood; but the congestion does not extend beyond the uterus, the Fallopian tubes, and the ovaries.*

These local phenomena are but the expression of a general law, which is dependent on the distribution of the ganglionic centres of the sympathetic system to every important organ of the animal body; and if an additional proof be required that independent nervous centres exist for the purpose of regulating the local distribution of blood, I would adduce the fact that, although local functions may be disturbed by injuring portions of the brain to which the motor and sensory nerves are affiliated, those local functions are generally restored long before the

* Pathological Transactions, vol. xiii., p. 170: Professor Harley on the Uterus and its Appendages at the Catamenial Period.

injured portions of the brain have had time to recover their integrity.

The localisation of such centres is obviously contingent on the position of the organs to which they are subservient, for such is the distribution of the ganglia in all animals from Echinodermata upwards; and each ganglionic centre, I suppose, gives nurture and individuality to its special out-going nerves, just as a modified soil may determine the quality of a sapling which has been protruded from the root of its parent tree.

From the foregoing considerations I conclude that the sensitive nerves of every ganglionic centre impart an influence, which I would call the vital sense of an organ, affecting its histological tissues, without operating immediately upon its bloodvessels; that the motor nerves, having their terminal fibres extended to the extreme ramifications of the arterial system, incite vascular action and secretion in response to the vital sense; and that the grey fibres of Remak, having a correlative ramification, restrain and regulate the stream of nutriment which is conveyed by the arterioles into the cell territory for secretion or assimilation.

These three forms of nerves are capable of adjusting the healthy balance of circulation so long as they remain undisturbed by accident or disease.

If I may be allowed to compare organic with inorganic things, I would represent the sensory nerve of every ganglionic centre as the steam of an engine, the ganglionic centre as the cylinder and piston rod, the motor nerve as the fly-wheel, and the fibre of Remak as the governors.

In concluding this part of my subject I will offer a few remarks on the pneumogastric nerves.

The nuclei from which the eighth pair of nerves derive their roots occupy the same tracts of grey nervous matter as do those from which every pair of spinal nerves originate ;*· and as their source is analogous so also is their function, but being distributed to very important organs, and very diversified opinions being entertained concerning them, I may be pardoned for considering them apart from all other compound nerves associated with the sympathetic system.

I need not dwell on the intercommunicating fibres between the several nuclei in the medulla oblongata, which associate the sensitive portions of the palate, the pharynx, the epiglottis, the larynx, and the respiratory mucous membrane, with the muscles of deglutition, of the voice, and of respiration; nor is it necessary to allude to the phenomena which attend either the section or stimulation of the upper branches of the pneumogastric, whether motor or sensory; the result of such experiments may be predicated. Of the vascular and cardiac branches, however, it is interesting to know that they are distributed exclusively on the surfaces of vessels, or round about the secreting cells of mucous surfaces.

Division of the pneumogastric nerves above the cardiac branches is always fatal, and in almost all cases death results from an excessive congestion of the lungs, the capillary vessels being found in a dilated condition, and a bloody serum exudes from

* Philosophical Transactions, vol. clviii., 1868 : On the Intimate Structure of the Spinal Cord and Brain. By J. L. Clarke.

them into the bronchial cells.* The secretion of mucus is arrested, as we are led to expect from corresponding results observed in all other cases, where the sensory nerve-fibres of a ganglion are destroyed, as those of the cardiac ganglion and of the minute pulmonary ganglia must be in such an operation.

It has been affirmed by some experimentalists that the division of the vagi paralyses the œsophagus, and arrests the secretion of gastric juice. The experiments of Waller on section of the vagus will explain how this may be the case;† but many careful observers, such as Breschet, Müller, Milne Edwards, Vavasseur, Holland, and Arnold, have asserted that the gastric juice continues to be secreted after section of the vagi.‡

An experiment of Bernard's, however, tends to explain the discrepancy, as well as to illustrate the fact that the vagi have an influence on the special vital sense of the stomach, and to associate its secretion with such influence as transmitted from the brain. He divided the two vagi of a dog which had a large fistulous opening through the parietes of the abdomen into the stomach, so that the interior of that organ could be seen. After the animal had fasted some time, when the stomach was empty, and its internal surface coated over with mucus, he removed the secretion by means of a soft sponge.

* "Leçons sur la Physiologie et la Pathologie du Système Nervaux." Par M. Claude Bernard, Tome ii. p. 520.

† Comptes Rendues: Waller on Section of the Spinal Roots of the Pneumogastric above the Inferior Ganglion, vol. xxxiv., p. 582-587.

‡ Dr. J. Reid: Cyclopædia of Anatomy and Physiology, vol. iii., p. 900.

No sooner was this done than the mucous membrane became red and turgid, and poured out large drops of gastric fluid, which trickled down its surface. At this moment the two vagi were divided in the neck, and immediately the membrane became pale, and the secretion—from being limpid and acid—became viscous, stringy, and alkaline.*

Notwithstanding all this, the ordinary supply of blood to the stomach appears to be regulated by the semilunar ganglia, the ultimate fibres of which extend to the minute arterioles, which have their middle coat entirely composed of muscular tissue.

These ganglia, and the coronary plexus of nerves proceeding from them, appear to be the expanded terminations of the splanchnic nerves, and the multipolar cells with which they communicate. The right vagus terminates in the left semilunar ganglion, and thus contributes, with the sympathetic, to form the isolating centre wherein the attributes of nerve-fibres are so modified as to subserve the local requirements of the stomach.

Guided by analogy, we may infer that nerve-fibres from the semilunar ganglion, after passing through the gastro-hepatic omentum into the transverse fissure of the liver to the coats of the vessels in the capsule of Glisson, whither I have traced them, terminate as other vaso-motor nerves do, and that the terminal fibrils of the pneumogastric or sensory nerves are distributed to the acini and biliary ducts.

It is in this cell territory that the ganglionic nerves adjust the balance of circulation, whereby, in

* "Leçons sur la Physiologie et la Pathologie du Système Nervaux." Tome ii. p. 417, 1858.

the healthy liver, the substance which Dr. Pavy has termed hematine is secreted. "This substance is prone to rapid transformation into sugar when in contact with complex nitrogenised animal materials ;"* and after death (the force or condition preventing sac-charine metamorphosis being lost) the transformation does take place. Now section of the splanchnic nerves induces an increased amount of sugar, and therefore an increased secretion of hematine.† Stimulation of the vagi has precisely the same effect.‡ But section of the vagi arrests the secretion of hematine, seeing that no trace of sugar in the liver is found after death.§

It is thus that the innervation of the blood-vessels of all the abdominal organs may be effected, and the heart itself may regulate the general circulation by the dictates communicated to it by reflex action from the sensitive fibres of the pneumogastsic nerves, and which are responded to by the motor fibres, or by the fibres of the sympathetic proper.

MM. Cyon and Ludwig have shown that it is through the splanchnics that these responses of the heart are communicated over the widest system of vessels in the body.‖

The sensitive nerve by which the reflex action of the heart is originated is a separate nerve in rabbits, and may be demonstrated as such. It is formed by two roots—one from the trunk of

---

* Dr. Pavy: Guy's Hospital Reports, vol. v. p. 204.
† Leçons sur la Physiologie et la Pathologie, &c., Tome ii. p. 529.
‡ Ibid., Exp. 21 Janvier, 1853, p. 144.
§ Ibid., p. 435.
‖ Cours Scientifiques, 1868, p. 421.

the pneumogastric, the other from the superior laryngeal nerve. It descends by the side of the carotid artery and sympathetic nerve, and on entering the thorax it communicates with branches from the first thoracic ganglion, and is eventually lost in the substance of the heart.

If this nerve be divided in the neck, irritation communicated to the lower cut end produces no effect ; but if the upper or centripetal cut end be irritated, a very sensible effect is produced on the blood-pressure in the carotid arteries.

By such a system of innervation the heart appears to be capable of regulating its own action according to circumstances, by exerting a reflex action on the important vaso-motor nerves of the general circulation.

The sensibility of the walls of the heart being excited by too large a supply of blood, reflex action is set up, which dilates the capillaries, diminishes the blood pressure, and causes the blood to accumulate in the periphery. If, on the contrary, the internal sensibility of the heart be too feebly excited, the peripheric vessels contract, and blood is attracted to the heart.

That ganglia exist in the substance of the heart itself was an *a priori* inference before Dr. Robert Lee demonstrated their existence; but that any of these ganglia are specially inhibitory or specially accelerating is, to my mind, an absurdity to suppose. In the comparatively membranous hearts of fishes and of some reptiles such ganglia exist, and the microscope reveals both grey and white fibres as connected with them. Each ganglion, therefore, is undoubtedly the centre of antagonistic nerve-fibres, and, consequently, of antagonistic actions.

The facts which I have thus strung together surely prove that in all cases, and in all organs, when the nerve-fibres of the sympathetic proper are in abeyance, and the motor nerve-fibres of the vaso-motor nerves are unchecked, hyperæmia of the capillaries is induced, and increased secretion the result; and *vice versâ* when the motor fibres of the vaso-motor nerves are in abeyance, and the nerve-fibres of Remak are unrestrained, anæmia of the capillaries is induced, and arrest of secretion the result.

On such positive data all other sciences have been established; but in medicine alone, as I have elsewhere shown, the human mind has been taxed to the utmost in constructing opinions, the conflicting nature of which has been but too often taken advantage of by the most dexterous charlatans; and it is with a view to assist in removing such reproach that I will utilise the foregoing facts for the purpose of showing that most remedial agents which have been used without an adequate appreciation of their special mode of action, may be employed with a positive knowledge of the tissues on which they act, and of their mode of action. And I venture to hope that my observations will be found to agree with the widest experience, and that they will commend themselves to the most practical good sense.

SECOND PART.

IF it be true, as I have attempted to show, that
every ganglionic centre of the sympathetic system
of nerves has three distinct elements over and above
the ganglionic cells, and that each element or nerve-
fibre has its own special attributes, I will further
endeavour to interpret, by the aid of the experiments and observations which I have adduced, the
modus operandi of many medicinal agents, and to
found upon the knowledge so acquired a rational
system of therapeutics.

Take, for instance, the fact shown by M. Claude
Bernard, that section of the sympathetic proper
induces increased vascularity and elevation of temperature in the parts to which those sympathetic
nerves are supplied, together with the fact that
there are certain medicines which have the faculty
of diminishing vascularity and lowering temperature, and we have at once some reason to suppose
that the agents in question have the power either
to subdue the force which is set free by the section
of the sympathetic, or to increase the force which
the sympathetic exercises as a restraining or inhibitory agent.

In some forms of inflammation—and inflammation, in some form or other, lies at the root of most
diseases—we have the pathological counterpart of

the result induced by section of the sympathetic—a state in which, owing to the loss of the inhibitory influence of the nerve-fibres of Remak, blood-corpuscles penetrate into those minute arterioles through which blood-plasma only should be propelled.

Now, in the spurred rye (ergota), we have an agent which produces the very opposite effect. By increasing the inhibitory influence of the nerves of Remak, it diminishes the calibre of vessels, and not only shuts out blood-corpuscles from the capillaries, but also the blood-plasma itself, and so restrains many forms of hæmorrhage. M. Brown-Séquard affirms that he has seen a manifest diminution in the calibre of blood-vessels in the pia mater of the spinal cord take place in dogs after they had swallowed large doses of ergota ; and in the convulsive ergotism which occurred in Germany in 1770 dry gangrene was a constant accompaniment.

Counter-irritants have a similar effect, but in a modified degree. M. Loven has shown that by irritating the peripheric extremities of sensitive nerves, a reflex influence is produced on the vaso-motor nerves, of an inhibitory character, effecting a diminution of the calibre of vessels.[*]

At a meeting of German naturalists and physicians held at Innsbrück in 1869, Professor Heidenhain, of Breslau, enunciated the fact, which he had observed from many experiments, that irritation of sensitive nerves produces a rapid diminution of the blood heat.[†] He also noticed a sensible decrease in the calibre of vessels, and a less frequent pulse.

[*] *British Medical Journal*, June 11th, 1870.
[†] Ibid,, Dec. 18th, 1869.

According to Professor Stilling, " there is a con.
stant reflex influence maintained by a sensitive
nerve upon the blood-vessel nerves." An excitant
applied to the extremities of a sensitive nerve
induces an increased flow of blood; but if the
extremities of a sensitive nerve be destroyed or
paralysed, a corresponding reflex action is induced
on the motor fibres of the vaso-motor nerves, the
ascendency is given to the inhibitory nerve-fibres of
Remak, and a diminished flow of blood ensues.

Schiff ascertained the fact that if the skin be
tickled or gently rubbed, the capillaries of the part
so excited dilate; but if great force be used in
rubbing, or if the extremities of the sensitive nerves
are greatly irritated, the capillaries do not dilate,
but firmly contract.* The strangury, for instance,
which is often occasioned by a blister applied to the
loins, is attended with a very scanty secretion.

It has been objected to counter-irritants that no
physiological explanation of their action can be
given; but M. Loven's demonstration of the effect
of irritation of the extremities of sensitive nerves
appears to me a good and valid explanation. The
observations which I have made on the so-called
trophic nerves, and the influence they have on the
vital attributes of parts, by increasing or diminishing the supply of blood, have a direct bearing on
this interesting subject.

Just as the localized ganglionic centres of innervation have an independent action, and accelerate
or retard the vascular function of any organ, so have
we medicinal agents which localize their power on

* " Year-book of Medicine and Surgery " for 1862, p. 19.

some special organ, and excite or restrain the functions of that organ, in a similar manner as does the section of the nerves of Remak, on the one hand, or the section of the motor fibres of the ganglionic centres, on the other. A delicate method of exemplifying the above proposition in a general way is by introducing into the blood poisonous substances which exert their action on special tissues, whilst, anatomically, those tissues maintain their integrity. Thus, we may suspend the vital functions of different portions of the nervous system. Anæsthetics, for instance, put a stop to sensation, whilst they leave motion intact. On the other hand, curare and aconitine suspend motion, whilst they leave sensation and volition intact.

We know that opium and chloral exert their action on the brain; digitalis and aconite on the heart; mercury and opium* on the liver; turpentine and uva-ursi on the kidneys; phosphorus and bromide of potassium on the organs of generation, etc.

Some of these drugs undoubtedly act on other parts of the body than those with which I have associated them; but for the purpose of suggesting a system of rational therapeutics I have ventured to refer to each as the type of a category in which others may be found to have a more exclusive, though perhaps less positive, action.

I have purposely named them in pairs, because I wish to show that each one has an antagonistic action to that with which I have conjoined it: thus,

---

* The *ulterior* or sedative influence of opium on the sensory nerves is here alluded to.

whilst opium stimulates the circulation, and produces an excitement of the motor nerves which often supersedes its hypnotic influence, chloral restrains the blood-current by acting as a powerful sedative to the motor and sensory nervous centres, and so suspends the activity which opium sets up. Opium accelerates the pulse, raises the temperature of the skin, increases the secretions of saliva and sweat, injects the conjunctivæ, and—according to Dr. J. Harley—if these manifestations of vascular excitement are prolonged, as they may be by repeating the medicine, they end in dilatation of the capillaries, general congestion, and imperfect oxidation of the blood; whereas chloral is a powerful sedative to the nervous system. Instead of accelerating the pulse, it reduces its frequency; instead of raising the temperature of the skin, it lowers it; and instead of increasing the secretion of saliva and sweat, it diminishes both.

I will quote one instance out of many already recorded, and from a source which may be the more readily accepted, because no theory is sought to be established in the description. In a case of puerperal mania treated by Dr. Head in the London Hospital, two drachms of the hydrate of chloral were given to the patient between 3 p.m. and 6 p.m., after which she slept twenty-four hours. Previous to the first dose her pulse was 120; her temperature 102·4°; and her respiration 30; after the chloral her pulse was 96; her temperature 99°, and her respiration 21; she complained, moreover, of great thirst.

Now, in cases of congestion of the brain, where stupor but wakefulness are the characteristic symptoms, surely it is more consistent with reason to give the chloral hydrate, or conium, than opium to procure sleep. I had a patient the subject of sunstroke, which occurred in Cuba in the summer of 1870. The gentleman came to England suffering from the effects in a state of mental hallucination, perfectly oblivious of having been to Cuba, feverish, and wakeful, and in a condition of constant restlessness. After several days, during which all the symptoms continued, two doses of chloral completely changed the whole aspect of the case. He fell into a state of almost constant sleep, but on waking had glimmerings of consciousness and memory; and, in short, somewhat to my surprise and that of Dr. Burrows, whose kind assistance I had, the patient eventually recovered.

Chloral, by diminishing the activity of the motor element of the vaso-motor nerves, appears to give the supremacy to the restraining or inhibitory fibres of Remak, and to diminish the calibre of vessels. The anæmic condition of the brain when under the influence of chloral may be demonstrated by the ophthalmoscope.

In an able address, read by Professor Laycock before the British Medical Association at Newcastle in 1869, it was shown that "no change in the mind or consciousness can or does take place without a coincident change in the brain;" "that the potential energy of the brain is the greatest of all kinds of vital energy, and *requires a larger supply of blood to maintain it;*" "that much more force as motion is used in mental than in bodily activity in the same time;" in other words, the larger the supply of blood the greater the energy of function. And in what diseases do we find the greatest amount of hyperæmia of the brain ? Why in cerebritis and mania, in which the physical results are precisely those induced artificially, by M. Claude Bernard, by section of the sympathetic in the neck, and, in a less degree, by opium. And what remedies have we which induce a directly opposite effect on the cerebral circulation? Chloral, as I have said, is one ; and the results of the tentatives, in the way of drugs, by physicians whose attention is directed to insanity exclusively, begin to show that on chloral will they mainly depend in that disease. Need I quote a better authority than that of Dr. Clouston, of Carlisle, who recommends to the Medico-Psychological Association the use of chloral, *to reduce the temperature, to lower the pulse, and to induce sleep ;* whilst (says he) opium and cannabis have an opposite effect.[*]

Conium, when taken in large doses, also acts as a sedative on the motor fibres of the vaso-motor nerves, and produces anæmia of the brain,[†] as do cicuta virosa, the aconitum napellus, and many of the Ranunculaceæ.

On the other hand, if we are satisfied that a sudden interruption of the nutritive supply of blood to the brain be the cause of an epileptic fit, as Kussmaul and Tenner have attempted to prove,[‡] then it would appear that opium or medicines producing

---

[*] *British and Foreign Medico-Chirurgical Review*, October, 1870.

[†] " Old Vegetable Neurotics " (Dr. J. Harley), p. 13.

[‡] Kussmaul and Tenner on Convulsions, New Sydenham Society, 1859.

directly or indirectly analogous results, should be combined with that which is known to prevent spasm of the glottis, on which the sudden arrest of blood to the brain may depend. I say *directly or indirectly* because I am satisfied that I have often found the addition of digitalis, which stimulates the heart, and induces an efficient contraction of the ventricles, with bromide of potassium, which prevents spasm of the glottis, effective in suspending epilepsy, when the bromide of potassium alone has failed. The oxide of silver, I think, acts precisely in the same way. I should here remark, however, that the ophthalmoscope, when used for determining the condition of the fundus of the eye, under the influence of the bromide of potassium, indicates an increased vascularity of the optic disc and retina ten minutes after the administration of half-a-drachm of that drug, and shows an increasing congestion during the lapse of several hours; hence, it is inferred that the bromide of potassium tends to produce congestion of the brain.*

M. Du-Bois Reymond attributes hemicrania to spasm of the muscular coats of the arterioles of the brain ; and Möllendorf recognises a form of anæmic hemicrania as resulting from irritation of the sympathetic, causing contraction of the minute vessels of the brain. And what medicines do we give to relieve such form of headache? Why, precisely those which impart an impetus to the cerebral circulation, as section of the sympathetic does,—stimulants and opium, and we also enjoin the recumbent

* " On the Action of certain Neurotics on the Cerebral Circulation." By Patrick Nicol, M.D., Physician to the Bradford Infirmary, and Isaac Mossop, L.R.C.P.E., of Edinburgh.

position, which has incidentally an influence in restoring a healthy blood current to the brain.

On such theory M. Bornetwick explains the curative action of ammonia and caffein, as stimulating the motor fibres of the vaso-motor nerves. And quinine, although it may stimulate the nerve-fibres of Remak, increases the blood-pressure, and sends an increased quantity of blood through the arterioles in a given time in such cases. Möllendorf also describes a form of congestive hemicrania, in which mild galvanism to the sympathetic in the neck is an effective remedy. I suppose the hydrate of chloral would be equally effective.

We have seen that section of the nerves of Remak which supply the parotid and submaxillary glands, is productive of an increase both in the circulation and secretion of those organs. Stimulation of the motor fibres by the administration of mercury superinduces the same phenomena. The antagonistic result, occasioned by section of the motor fibres, or by stimulation of the sympathetic proper, which go to these glands, is also induced by atropia. This latter alkaloid produces complete dryness of the tongue, roof of the mouth, and soft palate, extending more or less down the pharynx and larynx, rendering the voice husky, and often inducing dry cough and difficulty of deglutition, a parched state of the lips, and, occasionally, dryness of the mucous membranes of the nose and eyes.* That atropia localises its action in this way on the salivary glands and on the mucous surfaces of the pharynx and larynx, the epithelial structure of which belongs to the category of glands, is proved by the occur-

* "Old Vegetable Neurotics," pp. 203, 4.

rence of the same phenomena when atropia is injected into the subcutaneous tissue in any part of the body. We have, therefore, medicinal agents which stimulate or restrain the functions of the salivary glands, just as section or stimulation of the different nerves stimulates or restrains them; and I can imagine no other means of accounting for such medicinal action than its influence on the vaso-motor nerves.

When M. Claude Bernard divided the thoracic portion of the gangliated cord in the horse, he found that increased vascularity of the lungs and pericardium, together with serous exudation, were the results. Lobelia and digitalis also induce in-creased vascularity; but when the sympathetic nerve fibres retain their integrity, it would be unreasonable to expect the blood-stasis which generally results from section of those fibres. Chloral, acetate of lead, and, according to Mr. Blake,* the injection of the salts of soda into the blood, causing contraction of the pulmonary arteries, have the contrary effect.

I have a patient, a lady aged forty, who has occa-sionally attacks of asthma. Sometimes it assumes the dry form, when she is generally relieved by smoking stramonium until expectoration occurs. Sometimes the disease appears as *humoral asthma* of the old authors, when chloral relieves her quickly. She has once or twice taken chloral in the dry catarrhal attacks, and on each occasion—to use her own expression—has felt as though a tiger had hold of her throat. In the humoral form, if she smokes, the breathing is not relieved, but congestive head-ache follows for a day or two.

* *Edinburgh Medical and Surgical Journal,* vol. liv. pp. 341-346.

In bronchitis, with emphysema, where, in consequence of profuse secretion and impaired aëration of blood the nights are sleepless, chloral diminishes secretion, produces sleep, and promotes recovery; whilst opium will have the contrary effect. Cases illustrative of these facts are reported by Dr. Waters, of Liverpool.* Counter-irritation, on the principle I have quoted, as enunciated by MM. Loven and Hidenhain, has also a beneficial effect *in humoral asthma;* and Mr. Gaskoin has reported a case in point in the *British Medical Journal* (March 30, 1872).

According to Schiff, Rosenthal, and others, very feeble electric excitement of the vagus, below the inferior cervical ganglion, increases the force and frequency of the heart's action, probably by acting on the special cardiac accelerator nerve, which, as M. Cyon has shown, emerges from the spinal cord with the third branch of the inferior cervical ganglion; more violent excitement exhausts the irritability of the vagus, and stops the action of the heart.

Small and repeated doses of digitalis also increase the force and frequency of the heart's action. It may be objected that the physiological action of digitalis on the heart presents certain anomalies; and that some authors have ascribed to it an enervating influence, although others regard it as essentially stimulating. Dr. Sanders, of Edinburgh, was one of the first to affirm that, in small doses, digitalis occasions an increased action of the heart and febrile symptoms generally. Orfila endorsed that opinion, and Professor Christison has added the weight of his testimony to the same. In point of fact, it is now generally admitted that digitalis acts as a stimulus

* *Lancet*, May 4th, 1872.

to the heart; producing, where deficient power, irregular action, and dyspnœa exist, a more complete ventricular contraction, a firmer and more regular pulse, and relief from impeded respiration.* During the stimulating influence, secretion is increased, by virtue of the law which I have quoted, as dependent on the distribution of the ganglionic centres of the sympathetic system to every important organ. Where it is an object to enlarge the capillaries, and diminish the blood-pressure within them—and I have alluded to some cases of epilepsy as examples, —it seems but reasonable to give digitalis.

Excitation of the sympathetic trunk alone in the neck (and it may be done in rabbits, in which the vagus and sympathetic run separately) always causes increased blood-pressure, a lowering of temperature, but no alteration in the frequency of the cardiac beats;† and such are the effects of aconite. Dr. Wilks has given it in active inflammatory diseases, in acute rheumatism associated with endocarditis, and with satisfactory results. It appears to contract the arterioles, to prevent the accumulation of blood-corpuscles and their filtration through the capillary walls, to lower the temperature, and to abate the symptoms of fever generally. In short it stimulates the dormant activity of the fibres of Remak, and by so doing diminishes the calibre of

* Cases illustrative of such influence are given by Dr. B. D. Taylor, of the Bellevue Hospital, New York, in the New York Medical Journal for Nov., 1870; and Dr. Milner Fothergill, in the Hastings Prize Essay for 1870, has shown that digitalis is a member of a group of agents possessing such qualities. See also the British and Foreign Med.-Chir. Review, July, 1871: "Digitalis and Heart Diseases;" by Balthazar W. Foster, M.D., Professor of Medicine in Queen's College, Birmingham.

† Gmellin's Pflanzengifte, p. 717.

the arterioles. It is an interesting fact also that this drug appears to possess some of the attributes of a counter-irritant, for when chewed it produces intense tingling of the lips and tongue, and instances have occurred in which the leaves have blistered the skin. Dr. Fothergill has shown how that digitalis and aconite produce opposite action.*

I have described how the peculiar innervation of the heart renders that organ capable of regulating its own action, by exerting a reflex action on the widespread vaso-motor nerves of the general circulation.

Fever presents us with a pathological instance of the independent condition of the organs of circulation. With a contracted pupil there is a rise of temperature, the walls of the heart are preternaturally excited, the balance of function is suspended, the motor fibres of the vaso-motor nerves obtain the supremacy, and it is precisely those medicines which depress their function, or which stimulate the fibres of Remak, which are found to be most effectual as remedial agents. The alkaloid of veratrum, extolled as a remedy in fever by Trousseau and Aran in France, and by Vocher in Germany, has been carefully studied by Dr. Horatio Wood, of Philadelphia, who has determined that it exerts no direct influence on the brain; but that it depresses the functions of the spinal cord and heart, diminishes sensibility, restrains the action of the vaso-motor nerves, renders the respiration slower, and reduces the temperature of the body.†

But whilst its effect in lowering the temperature is conspicuous, the gastric and cardiac symptoms

---

* Hastings Prize Essay, 1870.
  American Journal of Medical Sciences, January, 1870.

which it evokes are sometimes so serious that we should give it with much circumspection, until we have data for predicating the class of cases in which these untoward effects are apt to be induced.

Aconitine and curarine have also the faculty of destroying the motor power of nerves, manifestly rendering those nerves incapable of conveying impression to the muscles; for the muscles themselves do not lose their contractility.*

We have the testimony of Dr. J. Harley that belladonna localises its action on the sympathetic nerves; that in moderate doses it induces a tonic and slightly contracted condition of the whole of the circulatory tubes, accompanied by increased force and frequency of the heart's action. "The blood," says he, "is equally distributed, and the circulation in any given part is so *tight* and rapid that it really contains a little less blood than when in a quiescent state, and the tissue is consequently a little paler. But the quantity which passes through it in a given time is generally increased."† After the use of moderate doses these effects are maintained throughout. Hence it suggests itself as a remedy in fever, and the more so as, being eliminated by the kidneys, it increases the secretion of urine. According to Dr. Meuriot, it also increases the secretion of bile, and predisposes to perspiration.‡ Mr. Christopher Heath, who has found it useful in fever, states that it reduces vascularity by diminishing the calibre of small vessels.

---

* MM. Gréhaut and Duquesnel, Comptes Rendus 73, p. 209.

† Dr. J. Harley; The Old Vegetable Neurotics.

‡ Dr. Meuriot; De la Méthode Physiologique à l'Étude de la Belladonna.

We have seen how it exercises a direct action on the salivary glands.

Quinine, beberia, and other antipyretic tonics appear to stimulate the nerves of Remak somewhat in the same manner. If it be true, as M. Binz has affirmed, that quinine prevents the formation of white corpuscules in the blood, which are never destined to arrive at the stage of red corpuscules—a process involving waste of force, accompanied with much deleterious heat; the increased blood pressure in the contracted arterioles may restrain the meta-morphosis of the blood globules, as supposed by M. Binz. The observations of Drs. Nicol and Mossop confirm the assumption that quinine diminishes the calibre of the arterioles.

Chloral, also, as a sedative to the sensory and motor nerves, re-acts on the vaso-motor nerves, lowers the temperature of the body, diminishes perspiration, and relieves the symptoms of fever generally, except that, like atropia, it does not allay thirst;* and that is perfectly consistent with its action generally of diminishing secretion. May not the apparent discrepancy in the fact that atropia in-creases abdominal secretion whilst chloral diminishes it, yet that both are beneficial in fever, be explained by the stimulating influence of atropia on the fibres of Remak and by the sedative action of chloral on the motor fibres of the vaso-motor nerves ? If so, we have remedies for fever generally, yet calculated to meet contrary exigencies in the course of that disease. It has been said that chloral was used in some of the French ambulances during the late war as a remedy against pyæmia, and this faculty, I apprehend, can depend only on the fact that, by deadening the influence of the motor fibres of the vaso-motor nerves, it virtually puts the fibres of Remak in a condition to occlude the minute vessels.

A volume might be written on the correlation of nerve and medicinal action upon the minute vessels of the stomach, and on the function of that organ generally; but the special influence of each does not appear to me to admit of the proof which I conceive to be adducible in relation to more exclusively glandular structures, because the evidence upon which any satisfactory inferences can be founded is so conflicting that I will not attempt to force the sway of fancy into the throne of truth. When men like Bernard, Kölliker, and Heidenhain, maintain one view respecting the influence of the pneumo-gastric and splanchnic nerves on the stomach, and Bidder, Vulpian, and Nasse entertained the very opposite, I may be excused from speculating on either side.

My conviction, however, is that the very same operation of medicinal agents on the respective fibres on the vaso-motor nerves which is localised, more or less, in other organs, obtains also in the stomach. Thus tartar emetic, taken into the stomach, injected into the veins, or even when rubbed into the palms of the hands,* produces redness and turgidity of the whole mucous surface of the stomach and intestines, and increased secretion precisely as does section of the splanchnic nerves or stimulation of the vagi, as performed by Bernard (see page 19); whereas lead, hydrocyanic acid, chloral, and opium, produce the same effect as does division of the vagi.

---

* Mem. of the London Medical Society, vol. ii. p. 386.

The influence possessed by the ganglionic nerve-centres on the chemical phenomena of life, can be simulated, increased, or diminished by special medicines. The increased supply of hematine in the cells of the liver, as caused by section of the sympathetic nerve-fibres, or by stimulating the pneumo-gastric nerves, may also be induced by mercury and iodine, and the very opposite effect attends the secondary action of opium, as I will presently explain.

And here I would adduce the liver to prove that something more than mere vascular fulness is necessary for the functional action of a secreting tissue; for such dilatation of vessels may, and often does, occur without a corresponding secretion of bile. By the suspension of the reflex influence of the sensory nerve-fibres, active hyperæmia of the hepatic artery may lead to hypertrophy of the organ, or hyperæmia of the portal veins may result in structural degeneration, without an increased outpouring of bile at any time.

Arguing on the experiments of Dr. Pavy and M. Claude Bernard (page 38), Drs. Devergie and Foville supposed that after section of the vagi the glycogenic function of the liver was paralyzed; that in diabetes they saw a vicarious function performed by the kidneys, and that by awakening the vaso-motor filaments of the sympathetic, and restoring the capillary circulation in the liver, diabetes should cease in many cases. Arsenic was employed to fulfil this purpose, and it succeeded. By parity of reasoning they explained the action of arsenic in intermittents. When taken in therapeutic doses it produces a rosy colour in the face, and increases the secretions generally, by which means it eliminates itself from the system.*

The very term *erythema mercuriale* is something like testimony that mercury acts on the sensitive nerves; and the more manifest physiological action of that drug is to produce a dilatation of the capillaries in a great number of important organs, and on a large extent of mucous surfaces. Orfila and Christison agree that in all cases of poisoning by mercury, when the first result is not chemical decomposition or corrosion of tissues, the leading symptoms are those of inflammation of the mouth, throat, and alimentary canal — salivation and mercurial erethism in one or more of its multifarious forms; bloody vomiting and purging, in conformity with Cohnheim's and Stricker's views of the permeability of the vascular walls by the blood-corpuscles.†

The influence of the splanchnic nerves and their ganglionic centres is distributed over a very wide area of the vascular system. Dr. Schmidt, of Rotterdam, associates Addison's disease with a morbid condition of the abdominal sympathetic; and there is great reason for the suggestion, seeing that the general anæmia, the palpitating heart, the soft and jerky pulse, and the feeling of faintness and breathlessness on attempting any exertion, are all indications of heart and vascular disturbance, with which Dr. Addison, assisted by Mr. Quekett, did find disease of the semilunar ganglion in one case of

* See M. Blachez, Gazette Hebdomadaire de Médecine et de Chirurgie, 1871.

† Ueber entzündung und eiterung. Virchow's Archiv, 1867, vol. ii.

leucocythæmia.* Professor Tigri, of Sienna, has also reported a case of hæmorrhage into the ganglia of the great sympathetic in which there was a bronzed skin and every other symptom of Addison's disease.†

The wide distribution of the influence of the splanchnics, as vaso-motor nerves, has its counterpart in the wide extent of the action of mercury and of iodine. Iodine, like mercury, produces dilatation of the capillaries generally, an exalted temperature, increased secretion, abundant menstrual discharge, priapism, and every indication of an excited circulation. The post-mortem appearances confirm those indications.‡

In gangrena senilis, every phase is an illustration of these views. We know that it may be produced by ergotine, and by the destruction of a compound spinal nerve. When it occurs spontaneously, every symptom may be interpreted by this theory of the vaso-motor nerves; from the first paroxysm of pain, which is followed by coldness of the part affected, through the consecutive stages of discolouration, loss of circulation, discharge, and mortification. Whilst writing these lines I am attending a case, and every impulse I have been able to give to the general circulation by stimulating the nerves of sensation and motion, the progress of the disease has been obviously retarded. Opium, iodine, and ammonia—all stimulants to the motor nerves—have given to the patient an almost superhuman power

* Addison's Works, published by the New Sydenham Society, vol. xxxvi. p. 213.

† Journal de Médecine de Chirurgie et de Pharmacologie. Bruxelles: November, 1870.

‡ Dr. Jahn, Archiv fur Medicinische Erfahrung, 1829, p. 342.

of resisting the disease; but no sooner was chloral given to induce sleep than every vestige of circulation was arrested; absolute mortification of the foot and leg set in, and ergotine very quickly induced the drying process.

In the catalogue of remedial agents, it is no uncommon thing to find some one which acts differently at different periods of time, or as the dose differs. Opium is one of these; its primary operation being that of a stimulant to the motor fibres of the vaso-motor nerves; but, ulteriorly, it is a sedative to both sensory and motor fibres; and thus it acts on those fibres of the hepatic plexus which, on entering the transverse fissure of the liver, are distributed, right and left, over the coats of the hepatic artery and portal veins. It has, therefore, an effect diametrically opposite to that of mercury. But I can well understand how calomel and opium may be combined to produce the incitive influence of the one with the calming effect of the other, so as to turn the balance of function on the side of the fibre of Remak, the attribute of which I have assumed to be to regulate the stream of blood which is conveyed by the arterioles into the cell territory for healthy secretion and nutrition.

This tranquillising influence is possessed by conium, and, especially so, by the bromide of potassium.

Mercury influences the vaso-motor nerves of the intestines, precisely as it does those of the liver. By stimulating the motor fibres, a preternatural secretion from the glandular and epithelia structures is induced, just as it is by the section of the coeliac and mesenteric plexuses, as performed by

Rudge on rabbits. Opium has an antagonistic action, as have also hydrocyanic acid, cannabis, and many astringents.

The *modus operandi* of diuretics generally is so different that we should have a clear apprehension as to what we propose to accomplish by their agency before we can expect to use them with success.

The term diuretic must be accorded to every therapeutic agent which increases the excretion of either the aqueous or the saline components of the urine; and when we use one of those agents it behoves us to consider which of the components it may be desirable to eliminate in greater quantity, and by what means to do it. In other words, on which of the three forms of nerve fibres do we propose acting.

Do we desire to increase the fluid portion only, either with the intent of carrying off from the system a quantity of water, or of flushing the tubuli uriniferi? Then those agents which stimulate the motor fibres of the vaso-motor nerves, and set up hyperæmia round about the Malpighian tufts, should be employed. Coffee, the action of which has been explained by M. Bornetwick, in hemicrania, scoparium, taraxacum, juniper, buchu, senega, and a host of others, bitters for instance, which we are in the habit of prescribing either in cases of general debility, attended with serous effusions, or when excess of the saline components of the urine induces irritability of the bladder.

Do we desire to increase the excretion of the components of the urine generally by exciting the whole secretory apparatus of the kidneys? Then should we depend on squill, colchicum, potash, turpentine, cantharides, mercury, iodine, arsenic, agents which stimulate the sensory nerve-fibres of the vaso-motor nerves, and thereby exalt the vital sense of the epithelial cells and of the organs generally.

On the other hand, if we desire to diminish the secretion of urine, we should either stimulate the restraining nerve fibres of Remak by belladonna, uva-ursi, ergota, etc., or subdue the relative power of the motor fibres, by conium, chloral, tobacco, bromide of potassium, etc., which are nervine sedatives to the sensory nerve fibres of the vaso-motor nerves.

Experiments on the nerves we must take with some degree of circumspection, because it has been found that certain lesions of the medulla oblongata, or section of the vagus or pneumogastric, may be productive of diabetes; but if diabetes be regarded as the result of disturbance of the functions of the brain and nervous system, or as cause rather than effect of kidney disease, for which we have high authority,* then may we reason on the correlation of nerve and medicinal action.

In nephritis we have the pathological counterpart of the experiment by Eckhardt (p. 20), except that the urine is scanty; but that may depend on the disease affecting the capillaries of the tubili uriniferi chiefly, and not on the Malpighian capillary system.

Turpentine and cantharides produce the same effects, even to the sympathetic vomiting, and pain in the testicle, thigh, and leg, by the specific irritation extending to the solar and spermatic plexuses. Belladonna, aconite, and uva-ursi, either by stimu-

* Neubaner and Vogel on the Urine; translated for the New Sydenham Society, by Dr. W. O. Markhem, p. 32.

lating the inhibitory nerves of Remak, or by sub-
duing the force of the sensory and motor fibres of
the renal plexus, have the very opposite influence.
I do not of course pretend to say that these drugs
should be depended upon exclusively in nephritis,
and that general or local depletion should be dis-
regarded as a means of relieving the congestive
turgor of the renal capillaries. My great object is
to illustrate the proposition with which I started,
namely, that as each element or nerve-fibre in the
ganglionic centre has its special attribute, either of
dilating or contracting the capillary vessels, so have
we medicinal agents which either stimulate the motor
nerve-fibres of the vaso-motor nerves, and dilate the
capillaries, or which act as sedatives on the motor
and sensory nerve-fibres of the vaso-motor nerves,
or give ascendency to the nerve-fibres of Remak,
and contract the capillaries.

If it be so, we have an explanation why the salts
of potash, which have been prescribed empirically
in nephritis, do more harm than good; for they
enter into the category of such medicines as juniper,
buchu, colchium, and squills, which stimulate the
motor fibres of the renal plexus, and increase the
secretion of the salts which characterise the urine.
We may also understand how it is that, in thus
acting as diuretics, they do not necessarily induce
diuresis. The additional stimulus which digitalis
gives to the general circulation by its action on the
heart may, I suppose, affect the capillary system of
the Malpighian bodies, and so flush the tubuli uriniferi and wash out the secretion of their epithelial
cells. Hence the great advantage of digitalis in
nephritis; for just in proportion to the outpouring

of fluid from the Malpighian capillary system, the turgescence of the tubular capillary system is diminished. Digitalis has been regarded almost as a specific by the Germans especially, and by Foville in France, in those cases of maniacal disease in which there is disproportioned strength in the carotid and temporal arteries, together with high-coloured and scanty urine.

In the straight tubuli uriniferi, which serve as conduits to the pelvis of the kidney, and in the ureters and bladder, we find the lamelliform epithelial cells which secrete mucus; and in most cases of preternatural mucous secretion from these parts, unconnected with mechanical causes, I can vouch for the beneficial action of ergota and bromide of potassium. In many forms of leucorrhœa, also, a combination of these drugs is very effective, when minerals, acids, and astringent injections fail.

These last remedial agents are nervine sedatives to the cerebro-spinal system, and consequently to the sensory and motor fibres of the vaso-motor nerves. As such, they lower the pulse, diminish the temperature of the body, and act as anaphrodisiacs on the organs of generation; whereas phosphorus, by exercising an antagonistic influence, is a powerful aphrodisiac. I need scarcely insist upon the inseparable association of this attribute, with an increased vascularity of the parts concerned.

From the facts which I have adduced, it may, I think, be inferred, and with moral certainty, that there is an invariable correspondence between functional activity and circulation; furthermore— and I have the authority of Schiff in asserting it— that the dilatation of the arterioles is neither the result of reaction after previous contraction, nor the mere result of an afflux of blood to an irritated part; nor is it independent of nervous action, but that it occurs under the influence of sensory and motor nerves.*

It can scarcely be objected to the proposition with which I started, that I have not adduced sufficient evidence to prove that vascular contraction is also the result of nervous influence; and it is impossible that one nerve, or one form of nerve, can perform the double office of dilating and contracting the calibre of vessels.

Seeing, therefore, that most pathological conditions are but modifications of physiological actions, and the effects of derangement of the operations of the vaso-motor nerves, on which the healthy functions of all organs depend, it appears to me that on the knowledge and due appreciation of such aids, we may found a rational and scientific system of therapeutics.

In these suggestions I do not, of course, include those remedial agents which act mechanically or chemically, but those only which have an influence on the vaso-motor nerves, and the great majority of medicines with which we are acquainted, will fall into the category of these latter.

There is every reason to believe also that they may be grouped in accordance with some natural principle, just as plants which belong to the same natural order possess similar characteristic properties, physiological and therapeutical.

Numerous instances could be given of this, and an interesting one is recorded by M. Narayan Dàji,

* Comptes Rendus, vol. lv. September 29th, 1862.

E

of Bombay, with whom Simaruba was a favourite
medicine, but, for the want of it, he anticipated
similar results from the Ailantus Excelsa (a tree
belonging to the natural order *Simarubaceæ*), the
stomachic properties of which more than realised
his expectation.*

An attempt has been made by Dr. Broadbent to
form chemico-therapeutical groups, and to show
that "substances closely allied chemically, must
have an analogous action on the system, or that the
diversity of their operation should be capable of
explanation on chemical principles." In other
words, that "chemical groups ought to form therapeutical groups."†

I have purposely abstained from alluding to alcohol as a remedial agent in inflammation, because of
the conflicting opinions which have resulted from
its employment. But if, by acting chemically on
the tissues generally, it restrains the metamorphosis
of the blood-corpuscles from rapid degeneration;
or, by inducing an increased blood pressure, it prevents the formation of an undue amount of white
corpuscles in the blood, which, as we have seen,
involves a waste of force, accompanied with mor-
bific heat; or, by rendering the walls of capillaries
less permeable to the blood plasma, it restrains the
albumen in the blood, which it is the nature of many
diseases to cause to escape, and most of these attributes it has been said to possess; then, I think, the
diffusible action of alcohol as a stimulant to the
body generally, acting probably on the motor-fibres
of the vaso-motor nerves, equalising the circulation,

* *British and Foreign Medico-Chirurgical Journal*, April, 1871, p. 520.

† Proceedings of the Royal Medical and Chirurgical Society, 1868,
p. 84.

and re-establishing the balance of function, which it is the nature of disease to disturb, must commend itself—*in medicinal doses*—to our reason, as it is said to have shown itself beneficial in practice.

In the treatise by Drs. Nicol and Mossop, to which I have referred, the ophthalmoscope showed the same result in all the cases after alcohol had been administered, namely, congestion of the optic disc, with the appearance of small vessels, not visible before, and congestion of the choroid and retina. On one occasion, a dose of two ounces of brandy was given forty-five minutes after giving a drachm of chloral, and the effect of the spirit in counteracting the anæmic condition brought about by the chloral, was exceedingly manifest. These observations would appear to endorse the opinion of Dr. L. Gros, that in pneumonia, and broncho-pneumonia of an adynamic character, alcohol diminishes delirium, induces sleep, lowers the pulse, and improves respiration.

It may be asked if the foregoing observations are likely to lead to an improved system of therapeutics? To which I would reply that were it only for the substitution of reason in the place of empiricism, even if practice remained unchanged, the gain would be great. If my theory be true, we may employ therapeutical agents with a more precise knowledge of their action, and therefore more effectively than formerly, although, in many cases, we may administer them to fulfil the same indications which simple observation has taught us they have the property of fulfilling.

Of this latter source of knowledge it was once said, that "if reason taught differently from experience, it was injurious; but if the same, it was superfluous." And yet the experience of centuries has not advanced the doctrine of therapeutics so far as to supply a satisfactory explanation of the *modus operandi* of any one remedial agent. I therefore venture to reason on the observations resulting from such experiments as I have recorded in this essay, and to maintain that they are sufficiently striking to justify the assumption that our system of therapeutics may be founded on a rational and scientific basis.

Evidence of facts is necessarily demanded in the satisfactory investigation of questions such as I have set forth, which are essentially questions of physical phenomena; and I appeal to the principle on which all knowledge is acquired, the credibility of the alleged facts, and the value of the testimony of those who have observed them, in justification of the conclusions which I have drawn from them.

Intrinsically the propositions have probability on their side; for if it can be shown that sections of different elements of the sympathetic ganglia affect the functions of organs differently, by increasing or diminishing their actions, and that we possess medicinal agents which produce analogous effects, the inevitable inference is that in all such remedies we possess just so many equivalents of what is virtually nerve force.

The instances I have given may seem insufficient as a basis for such an important generalisation; but I have tested my opinions in practice for many years, and have no reason to be dissatisfied with the result.

London: Pardon & Son, Printers, Paternoster Row.